Gob the Gnome

Gob and the Return of the Gnomes

Written By:

Marilyn Slaughter, Ed.D.

Illustrated By:

Angel Neha

Gob the Gnome™ Gob and the Return of the Gnomes

Copyright © 2022 by Marilyn Slaughter, Ed.D. All rights reserved.

Published by Majestic Children's Books, LLC

2817 West End Avenue | Suite 126-212| Nashville, TN 37203 | USA

www.gobthegnome.com

Printed in the United States of America

ISBN:979-8-9900545-2-3

Finally, the storm stopped. It was a horrible experience for Iris and Nathan's family. Feeling afraid and unsure of the aftermath of the heavy winds blowing all day was a lot for them to handle. The direction of the wind blew away from Iris and Nathan's home and towards the forest. They were concerned about their friends in the forest. Gob, Nancy, Koemi, Corni, and Mera had no where to shelter themselves from the storm. Nathan and Iris wanted to go to the park to check on their gnome friends after the storm. Nathan held his sister's Iris hand tightly as their mother drove them to the forest. The children saw the devastation of the storm in neighboring areas. Some houses were severely damaged; broken windows, leaking roofs, and downed powerlines everywhere. The storm had shown no mercy for anything in its path.

When Iris, Nathan and their mom pulled up to the parking lot Ranger Lisa was already waiting for them. Mom had called ahead to let Ranger Lisa know they were on their way to the park. As soon as the car stopped Nathan and Iris lunged outside and ran towards Ranger Lisa. They wrapped their arms tightly around Ranger Lisa, with a grasp of love and concern for her safety.

"Everything is alright," Ranger Lisa said as she hugged Nathan and Lisa. "Nobody got hurt from the storm, thankfully."

Nathan and Iris, and their mom felt relieved that none of their friends were injured in the storm.

"There is a lot of damage to the forest," Ranger Lisa warned them. "Let us find Gob and Koemi. They are over at the lake to see if Mera is okay," Ranger Lisa said.

Nathan, Iris, and mom followed Ranger Lisa to the water's edge. Mera did not often leave the water. She was sitting on the sand close to Gob, Nancy, Corni, Herb, and Koemi as they were trying to comfort their friend.

"Gob," Nathan called out softly to alert the gnomes of their presence. Gob looked up with a glossy look in his eyes. It was obvious that he had been crying. One look around them was enough to know that the gnomes and animals had endured a great deal of trauma from the affects of the storm. There were trees broken, bent, and completely uprooted as far as your eyes could see.

Everyone hugged each other and waited patiently for Gob to tell them about the damage in the park that was not in plain sight.

Environment

"It was horrible," Gob said. "When the storm started, I thought it would just be rain. Nothing to worry about. I was very wrong."

Koemi nodded: "It turned bad very quickly."

"First, branches started falling from the trees. Then, whole trees came falling down along with the critters that lived inside them. The fallen trees crushed the homes of other animals, too. This is the biggest disaster I have experienced in my entire life. I do not know how I am going to begin to fix the damage to the forest," said God.

Gob rubbed at his eyes to prevent the tears from falling.

"We'll help you," Iris promised as she placed a hand on Gob's slumped over shoulder.

"Kids your kind offer is nice but I need even more help to clean up the disaster. The forest is too big for just nine of us to repair and salvage the surroundings," said Gob.

13

"I wish our friends were here," Corni said with a long face. Corni was the eldest of the gnomes and he longed for the good old days when gnomes ruled the forests. Since Gob revived Corni from being a stone lawn ornament, his memory takes him back to long ago. Corni's memory is slowly returning to current time.

"I remember when this forest was full of gnomes," Corni said. "We were sent here by royalty from all over the world thousands of years ago. They needed a new place to hide their valuables from village raiders and demons. Royals and wealthy people of the period trusted gnomes to protect and secure their valuables within the vast pristine land of the unknown," he said.

Environment

"For many years it was only the gnomes, animals and trees on this beautiful land. The gnomes during that period were valued for their excellent talents, strength, and ability to survive off the land," said Corni.

"Why did you leave the forest," Iris asked curiously.

Environment

"We left to contribute our skills and knowledge about the environment to help human's progress in their new society. We embraced the modern way of life and thought our wisdom would be accepted," Koemi said. "Humans misunderstood the gnomes who made gestures of hope and prosperity for a healthy environment," Corni added. "We thought being in a smaller park we could educate several groups of people about protecting the environment and ecosystem surroundings. We hoped these people would share what they learned with friends and family," said Gob. We wanted to spread our strengths and understanding about protecting the environment as we did during ancient civilization and how to live safely and in harmony with nature," Gob continued. "The modern world offered people so much that the communities we sought out to help had other concerns and interests," God said sadly.

"Some people are still willing to learn and help the environment," Nathan protested.

Koemi rubbed Nathan's back, "Yes, you and your sister have given us new hope for the future."

"Well, I think that hope has been sufficiently dashed by this storm," Herb's face showed sadness as he looked around at the destruction surrounding them.

"We'll do whatever we can to help clean up the park, right mom," Nathan said. Simultaneously, Nathan and Iris turned to their mother who nodded in agreement and said, "yes."

"Thank you, but I'm afraid that won't be enough this time," Gob said.

On the drive back home Nathan and Iris were so quiet that their mom hardly recognized her own kids. She had never seen them so deep into their thoughts before. Even her suggestion of stopping for a burger and fries was met with silence.

"I know you kids are worried," mom said, "but I'm sure that Gob and the others will get the park back in shape again soon."

"The park is too big to care for with only five Gnomes," Iris said.

Nathan nodded: "I bet a lot of people in our neighborhood still have old lawn gnomes stuffed in a box somewhere in their garage."

"If we return them to the forest, Gob can revive the gnomes back to life," Nathan said.

"But how will we convince the neighbors to give them to us?" asked Iris.

The children thought about an answer for a while.

"Maybe we can convince our neighbors to donate their gnomes to the community garden," mom suggested.

"Then, what will we say to our neighbors when they see that their gnomes are all gone?" asked Iris.

Mom had not thought about that, but Iris was right, mom thought to herself. "We would not be able to explain the gnome's disappearance," mom whispered to herself.

"I don't think there's an easy solution for this problem kids," mom said after another couple of minutes went by in silence. "Maybe after a nice dinner and a good night's sleep, we will wake up with some fresh ideas that will work out this situation," mom said.

"Gob does not have time for us to figure out a solution overnight. Animals have lost their homes because of the storm," said Nathan with concern.

Mom sighed. Finally, they agreed to think about it for the rest of the day and discuss any ideas they produce during dinner. Nathan left home as soon as they arrived to explain the dire situation to Tucker. Nathan had a clever idea that he wanted to discuss with Tucker first, before telling mom and Iris.

227

ris went to the community garden to clear her head and water some of the flowers. Mr. Henderson was there and was planting a beautiful pink flower that made her think of princesses and fairies.

"Hi, Mr. Henderson! What is that?" asked Iris.

The old man greeted Iris with a warm smile and said, "It is called snapdragon and bees love them. I thought I would plant the snapdragons because these beauties do well in colder climates."

"They're very pretty," Iris said as she leaned in to admire the flowers. Which gave Mr. Henderson a good opportunity to notice the young girl's worried look.

"Are you alright," Mr. Henderson asked with a concerned look on his face.

"Mom took us to Sunbrook Forest today," Iris answered. "It's a wreck," Iris sobbed as she told her neighbor, Mr. Henderson. Mr. Henderson became a good friend of Iris' family ever since they settled their differences about the community garden. "There are fallen and damaged trees as well as destruction to nests, holes, and habitats of the animals who live in the forest," Iris said sobbing.

Community
Garden
25

"Now we need a lot of gnomes to fix everything." said Iris.

"Gnomes? Why do you need gnomes to help clean up the forest?" asked Mr. Henderson.

Realizing she had said too much, Iris bit her lip trying to produce a solid explanation. Iris remembered Corni's speech.

"Well, long ago kings and queens had gnomes look after their forests, crops, and treasures. They brought stability and good luck to the people and forest as well. I just think that the forest could use a little bit of luck after all that has happened," Iris explained.

Mr. Henderson thought about Iris's words. He had a pair of gnomes in his own garage and he would wager other neighbors had them as well. Of course, Mr. Henderson had not made himself popular in the neighborhood the last few years. Ever since his wife passed away, Mr. Henderson had been known as the grumpy old man. He liked to rat out the hiding places of kids playing hide and seek on their street. Mr. Henderson was opposed to the community garden in the neighborhood and thought it would only be a place for children to skateboard. He did not sign the petition and convinced other neighbors not to sign the petition too.

Community
Garden
27

Mr. Henderson told Iris that if she believed that gnomes would help fix the park, he would find a way to fill the forest with them. Mr. Henderson believed in the work Iris and Nathan were doing with the forest and community garden. He thought to himself that this little girl could achieve anything she set her mind to, and her brother was just as committed to community service.

Just when Mr. Henderson thought of Nathan, Iris' older brother came running over to inform his sister of dinner. Nathan came at the right time to walk Iris home, a plan had started to formed in Mr. Henderson's head, and he needed to be alone to set it in motion.

A week passed, Iris and Nathan were eager to return to the Sunbrook Forest. Every time they asked their mother to drive them to the park, their mom had an excuse as to why they could not go that day. Iris and Nathan chatted among themselves. They did not understand why they could not go to the forest park to visit with their friends and have fun. It was summer vacation and there was no school. Nathan and Iris thought their mother would be relieved to see them so eager to help clean up the forest. Their mom has always wanted her children to have balance in their lives. Spend some time outdoors and some time at home enjoying their computer games and streaming movies.

It was early on Friday morning when the doorbell rang. Mom, Nathan, and Iris were still sitting around the breakfast table. Mom stood up to see who was at the door. She returned followed by Mr. Henderson who had a big smile on his face that seemed to stretch from ear to ear. Iris and Nathan exchanged a glance, they had never seen their neighbor so happy before.

"Good morning Mr. Henderson," Iris and Nathan greeted him. Confusion was written on both of their faces. Nathan was the first one to notice that their mom was also grinning. Whatever was going on, mom was in on the scheme the children believed.

"What's going on?" Nathan asked, confused.

"You kids will find out soon enough," mom said. "Finish your breakfast first; then, we are going for a drive," she said.

ris looked at Mr. Henderson. "All four of us?" Iris asked.

"Yep," Mr. Henderson said with a wink.

Mr. Henderson looked more like a kid about to spend his whole allowance in a candy shop than an old man. As soon as the car pulled up to the park, Mr. Henderson jumped out and opened the door for Iris. Nathan hopped out on his own on the other side of the car.

"Why are we here," Iris asked.

Mom answered, "Mr. Henderson has been working very hard to organize a surprise for you this past week."

he old man nodded while Iris' mom took out her phone. Nathan and Iris huddled close to be able to see the screen.

"When we last spoke at the community garden you told me that you thought the forest needed more gnomes because they bring stability and good luck," said Mr. Henderson.

Iris cringed as she felt her brother Nathan look at Mr. Henderson strangely. Iris wished she could explain to Nathan that she had not revealed too much. Mr. Henderson was standing beside them and it was impossible for Iris to say anything to Nathan about what happened.

"So, I remembered when your video about the community garden was infected," said Mr. Henderson.

"Went viral," mom corrected the old man.

"So, I made a video of my own and it went......" Mr. Henderson turned to mom for help.

"Viral."

"I am a Clock App celebrity now," he said with a grin. "Play the video please, Miriam."

Mom pressed the play button.

Mr. Henderson was looking straight into the camera like a deer in headlights. Nervously, he mentioned how Nathan and Iris were the most inspiring young people he had ever met. Their love for nature and the environment is astonishing to me. Next, he talked about how the storm had affected the forest and how Iris had been heartbroken. As the video continued, Mr. Henderson shared with the viewers Iris' story about the gnomes. The final part of the video showed pictures from a park in the Netherlands called a 'Kabouter pad.' An attraction in the national parks where children and their parents can spot gnomes along a hiking path."

Community
Garden
39

"I have asked everybody who watched the video to send me their gnomes to create a 'Kabouter pad' in our national park. This week alone the mail carrier delivered over thirty gnomes to my house. I got another seven gnomes from neighbors. I created the little doors for the gnome's houses myself. Then, I asked for permission from the park rangers to set out a trail for the gnome's little houses. The Rangers told me not to permanently alter the forest when making the trail," Mr. Henderson explained.

"What?" Nathan stammered as he squeezed his sister's hand with excitement.

"I've created a trail in the forest that takes hikers from one gnome-house to the next," explained Mr. Henderson.

41

Nathan and Iris could not believe their ears. Sure, the neighborhood grouch had been kinder since he got involved with the community garden. The children could never imagine an act of kindness on this scale. Iris and Nathan threw themselves at their neighbor and hugged him tightly. "Thank you so much Mr. Henderson," Nathan said misty eyed.

"You are the best neighbor," Iris added.

"Come on, let's check out the trail," Mr. Henderson said, feeling a bit embarrassed by the praise.

They spotted the first gnome house very quickly; it was just a small door fixed to a tree trunk. Seeing the first tree trunk made Iris squeal in delight. Mr. Henderson scratched his head and said, "I would've sworn I left a gnome in front of it."

Not thinking much of it they continued their path. Mom spotted the second door but again the gnome Mr. Henderson had left at the door was gone.

he same thing happened at the third house;
then, the fourth house, and all the other houses
Mr. Henderson had so carefully crafted and
installed.

"I don't get it," Mr. Henderson said as they
neared the end of the route as well as the lake. "I left
this forest swarming with gnomes when I left
yesterday. I cannot believe someone took them all
away."

"They didn't," a voice called out from the tree
line.

Everybody turned around and gasped as they saw
dozens of little men and women approach them. Mr.
Henderson clasped his hand over his chest to steady
his rapidly beating heart. Iris grabbed his free hand.

Environment

"Do not worry Mr. Henderson, this is our friend Gob. He is a Gnome. He does not typically reveal himself to strangers," said Iris.

Mr. Henderson nodded, not feeling relieved by Iris' words in the slightest.

"We make an exception for people with a good heart," Gob said as he walked over to Mr. Henderson and reached out to shake his hand. Mr. Henderson shook Gob's hand tentatively. "You have returned my friends to their home, and I returned them to life. I will be forever grateful to you," Gob swore.

I ris, Nathan, and mom stared at the small army of
gnomes who all looked so grateful and happy to be
home again at last. Behind the line of gnomes
Ranger Lisa followed. The crowd split to let her pass
to the front.

"I still can't believe this," Ranger Lisa said teary-
eyed. "This is a miracle," she said. "The gnomes have
worked so hard already to fix the damage caused by
the storm. Did you know they all have specialties?
Ladle has tended to the wounded trees, healing their
wounds. Posy has made strong potions for the
animals to help them regain their strength.
Nibleclaw is an earth whisperer; he helped animals
who lost their burrows to find the best places to build
new ones. There is just too much to mention,"
Ranger Lisa with a big smile on her face.

ris and Nathan looked up as Gob called their names. The kids walked over to their gnome friend, Gob, and kneeled beside him. Their mom and Ranger Lisa ushered everybody else away except for Koemi, and Mera, who had just popped up from the lake and was pulling herself onto the shore.

Iris smiled: "Now that your friends are back you can restore the forest to its full glory," she said.

"You probably won't need our help anymore," Nathan thought aloud as he glanced over his shoulder at the gnomes behind them.

"Are you kidding," Gob asked, "You kids have done so much for me and this forest. I will still need your help and your presence in the park. You are my friends, aren't you?

"Yes," both kids agreed rapidly.

"**Y**ou helped me get rid of people who tried to cut down my precious forest illegally, you initiated a clean-up crew to help keep this place clean and safe for the animals, and you have made bees return to the woods. You kids are marvels and I consider the day we met to be the luckiest day of my life!" Gob said excitedly.

Nathan wrapped his arm around his sister's shoulders knowing that she was close to tears.

"I just wanted to tell you both," Gob held out his hands and both kids took hold of one, "thank you both from the bottom of my heart."

"Thank you," Nancy added.

"Thank you so much," Mera said.

They spent a wonderful day in the forest helping the gnomes fix the damage from the storm. Iris told Mr. Henderson all about their first encounter with Gob. Not surprisingly, Gob and Mr. Henderson were friends almost immediately. At the end of the day the gnomes returned to their homes and so did Nathan, Iris, mom, Ranger Lisa, and Mr. Henderson.

The adventures of Gob the gnome, Nathan and Iris continues as the children move into middle school.

THE END

56

57

Glossary

Clink – A sharp ringing sound, such as that made when metal and glass are struck together.

Condensation – The change of the state of matter from gas to liquid. Cork – the buoyant, light brown substance obtained from the outer layer of the bark cork oak.

Cork – The buoyant, light brown substance obtained from the outer layer of the bark cork oak.

Disintegrates – Break down into parts; to fall apart; reduce to pieces.

Documentary – Consisting of official pieces of written, printed or other material; factual film.

Drought - A period of dry weather, especially a long one that is injurious to crops.

Dumped – To throw away; abandoned.

Ecosystem – A place where plants and animals (including humans) live in relationship with each other and with the surrounding physical conditions.

Evaporates – Turn from liquid into vapor.

Expression - A particular word, phrase, or form of words.

Floppy - Limp or hanging loosely. Gap - A break or opening, as in a fence, or wall.

Hydrated – Having the right amount of water inside and outside of each cell in the body.

Keeled – Be in or assume a position in which the body is supported by a knee, as when praying or showing submissions.

Landscape – All the visible features of an area of countryside or land.

Littering - Objects strewn or scattered about; scattered rubbish; a condition of disorder or untidiness.

Manufacturer - a person, group, or company that owns or runs a manufacturing plant.

Meekly – In a quiet, gentle, submissive manner.

Photosynthesis - The complex process by which carbon dioxide, water, and certain inorganic salts are converted into carbohydrates by green plants, algae, and certain bacteria, using energy from the sun and chlorophyll.

Plucked - To pull off or out from the place of growth, as fruit, flowers, feathers, etc.

Recycle - To treat or process (used or waste materials) to make suitable for reuse.

Smoldering – Small, smoky burning fire.

Trickles – When a liquid flows in a small stream.

Wail – A prolonged high-pitched cry of pain, grief, or anger.

Whimpered - When a person or animal makes a series of low, feeble sounds expressive of fear, pain or discomfort.

Wrenching – To pull or twist (someone or something) suddenly and violently.